META - CALCULUS :

DIFFERENTIAL AND INTEGRAL

Jane Grossman

University of Lowell
and
Archimedes Foundation

Archimedes Foundation
Box 240, Rockport
Massachusetts 01966

ISBN 0977117022

"for each successive class of phenomena, a new calculus or a new geometry, as the case might be, which might prove not wholly inadequate to the subtlety of nature."

Quoted, without citation,
by H. J. S. Smith; *Nature*,
Volume 8 (1873), page 450.

First Printing, July 1981

PREFACE

In this monograph I present some new systems of calculus, which are called meta-calculi because in certain ways they transcend the classical calculus developed by Newton, Leibniz, and their predecessors.

The differential branch of the classical calculus is based on the concept of the average rate of change of a function f on an interval [r,s], which may be defined as

$$\frac{f(s) - f(r)}{s - r},$$

and will be called a classical gradient. Clearly it depends only on the endpoints (r, f(r)) and (s, f(s)), and it is independent of all the intermediate points (x, f(x)), where $r < x < s$.

On the other hand, the gradients that give rise to the meta-calculi depend on ALL the intermediate points and on the endpoints. Such gradients, here called meta-gradients, were first defined by my husband Michael Grossman and me, using an idea suggested by our colleague Robert Katz.

The definition of a meta-gradient depends on the selection of two weight functions, which are applied to function arguments and values in special ways. By means of the usual limit process, the meta-gradients then lead to meta-derivatives, which provide instantaneous rates of change that differ from the classical.

For each meta-derivative there is a corresponding meta-integral; and their relationship is given by two fundamental theorems, one of which resembles the Second Fundamental Theorem of Classical Calculus, and one of which differs in form from the First Fundamental Theorem of Classical Calculus.

In Chapter One, I present some basic ideas of classical calculus in a manner that leads naturally to the development of the meta-calculi in Chapter Two.

Since this self-contained work is intended for a wide audience, I have included many details that would not appear in a research report. And I have excluded proofs of all theorems, most of which can be established in a straightforward way. It is assumed, of course, that the reader has a working knowledge of classical calculus.

Finally, I wish to thank Michael Grossman and Robert Katz for their assistance, which was essential to the final development of the theory presented here.

The meta-calculi arose from the problem of measuring stock-price performance when taking all intermediate prices into consideration.

Jane Grossman

185 Florence Road

Lowell, MA 01851

July 14, 1981

CONTENTS

PRELIMINARIES

The word number means real number. The letter R stands for the set of all numbers.

If $r < s$, then the interval $[r,s]$ is the set of all numbers x such that $r \leq x \leq s$. (Only such intervals are used here.) The interior of $[r,s]$ consists of all numbers x such that $r < x < s$. The classical measure of $[r,s]$ is $s - r$. A unit interval is any interval whose classical measure is 1.

An arithmetic partition of $[r,s]$ is any arithmetic progression whose first term is r and last term is s. An arithmetic partition with exactly n terms is said to be n-fold.

A point is any ordered pair of numbers, each of which is called a coordinate of the point. A function is any set of points, each distinct two of which have distinct first coordinates.

The domain of a function is the set of all its arguments (first coordinates); the range of a function is the set of all its values (second coordinates). A function is said to be on its domain and to be defined at each of its arguments.

The classical change of a function f on $[r,s]$ is $f(s) - f(r)$.

CHAPTER ONE

Classical Calculus

1.1 INTRODUCTION

In this chapter some basic ideas of classical calculus are presented in a manner that leads naturally to the development of the meta-calculi in Chapter Two.

1.2 LINEAR FUNCTIONS

A linear function is any function that is on R and is expressible in the form bx + c, where b and c are constants.

Clearly the classical change of each linear function is the same on any two intervals with the same classical measure. In view of that property each linear function is said to be classically uniform.

Of course, each linear function has the same classical change on every unit interval.

1.3 CLASSICAL SLOPE OF A LINEAR FUNCTION

If ℓ is a linear function, then on EVERY interval [r,s] the ratio

$$\frac{\ell(s) - \ell(r)}{s - r}$$

has the same value, which we call the <u>classical slope of ℓ</u>.

Clearly the classical slope of a linear function is equal to its classical change on every unit interval.

The classical slope of the linear function bx + c is b.

1.4 CLASSICAL GRADIENT

The differential branch of classical calculus is rooted in the average rate of change, which we prefer to call the classical gradient. (The term "gradient" is better suited for our purpose, since it can readily be modified by appropriate adjectives. Of course, "gradient" is also used in vector analysis, but that subject does not concern us here.)

The <u>classical gradient of a function f on [r,s]</u> is denoted by $G_r^s f$ and is defined to be the ratio of the classical change of f on [r,s] to the classical measure of [r,s]; that is,

$$G_r^s f = \frac{f(s) - f(r)}{s - r}.$$

The operator G is

<u>Additive</u>: $G_r^s(f + g) = G_r^s f + G_r^s g,$

<u>Homogeneous</u>: $G_r^s(k \cdot f) = k \cdot G_r^s f,$ k any constant.

If ℓ is a linear function, then $G_r^s \ell$ equals the classical slope of ℓ.

Finally, when r = s, the expression for the classical gradient yields the indeterminate form 0/0, which brings us to the next topic.

1.5 CLASSICAL DERIVATIVE

Let f be a function defined on an interval containing the number a in its interior. Notice that for each number x in that interval,

$$\frac{f(x) - f(a)}{x - a} = \begin{cases} G_a^x f & \text{if } a < x, \\ G_x^a f & \text{if } x < a. \end{cases}$$

We now define the <u>classical derivative of f at a</u> to be the following limit, if it exists,

$$\lim_{x \to a} \frac{f(x) - f(a)}{x - a},$$

and we denote that limit by [Df](a).

The classical derivative of f, denoted by Df, is the function that assigns to each number t the number [Df](t), if it is defined.

The operator D is

Additive: $[D(f + g)](a) = [Df](a) + [Dg](a)$,

Homogeneous: $[D(k \cdot f)](a) = k \cdot [Df](a)$, k any constant.

Of course, if f is a constant function on R, then Df is everywhere equal to 0.

Furthermore, the classical derivative of each linear function has a constant value equal to its classical slope. Indeed, only linear functions have classical derivatives that are constant on R.

We conclude this section with a definition, which will be useful in Chapter Two. A smooth function on [r,s] is any function f such that Df is continuous on [r,s].

1.6 ARITHMETIC AVERAGE

Let f be any continuous function on [r,s]. Then the arithmetic average of f on [r,s] is denoted by $A_r^s f$ and is defined to be the limit of the convergent sequence whose nth term is the arithmetic average

$$\frac{f(a_1) + \cdots + f(a_n)}{n},$$

where $a_1, \ldots, a_n$ is the n-fold arithmetic partition of [r,s].

The operator A is

<u>Additive</u>: $A_r^s(f + g) = A_r^s f + A_r^s g$,

<u>Homogeneous</u>: $A_r^s(k \cdot f) = k \cdot A_r^s f$, k any constant.

The operator A is characterized by the following three properties. (This use of the term "characterized" indicates that no other operator possesses all three properties.)

1. For any interval [r,s] and any constant function k(x) = c on [r,s],

$$A_r^s k = c.$$

2. For any interval [r,s] and any continuous functions f and g on [r,s], if $f(x) \le g(x)$ on [r,s], then

$$A_r^s f \le A_r^s g.$$

3. For any numbers r, s, t such that $r < s < t$ and any continuous function f on [r,t],

$$(s - r) \cdot A_r^s f + (t - s) \cdot A_s^t f = (t - r) \cdot A_r^t f.$$

1.7 THE BASIC THEOREM OF CLASSICAL CALCULUS

First we state a simple result about any function h on [r,s].

Let $a_1,...,a_n$ be any arithmetic partition of [r,s]. Then

$$\frac{G_{a_1}^{a_2}h + \cdots + G_{a_{n-1}}^{a_n}h}{n-1} = G_r^s h;$$

that is, the arithmetic average of the n-1 successive classical gradients of h determined by $a_1,...,a_n$ is equal to the classical gradient of h on [r,s].

The preceding result suggests the following important theorem.

Basic Theorem of Classical Calculus

If Dh is continuous on [r,s], then its arithmetic average on [r,s] is equal to the classical gradient of h on [r,s]; that is,

$$A_r^s(Dh) = G_r^s h.$$

In view of this theorem we say that the arithmetic average fits naturally into the scheme of classical calculus.

1.8 THE BASIC PROBLEM OF CLASSICAL CALCULUS

Suppose that f, the classical derivative of h, is continuous and known at each number in [r,s]. Find h(s) - h(r).

Solution

By the Basic Theorem of Classical Calculus,

$$A_r^s f = G_r^s h = \frac{h(s) - h(r)}{s - r}.$$

Hence,

$$h(s) - h(r) = (s - r) \cdot A_r^s f.$$

The number $(s - r) \cdot A_r^s f$ arises with sufficient frequency to warrant a special name, "the classical integral of f on [r,s]," which is discussed in the next section.

Thus, the Basic Theorem of Classical Calculus, which involves the arithmetic average, classical derivative, and classical gradient, provides for the Basic Problem of Classical Calculus an immediate solution, which, in turn, motivates our definition of the classical integral in terms of the arithmetic average.

1.9 CLASSICAL INTEGRAL

Let f be any continuous function on [r,s]. Then the <u>classical integral of f on [r,s]</u> is denoted by $\int_r^s f$ and is defined to be the number

$$(s - r) \cdot A_r^s f.$$

We set $\int_r^r f = 0$ and $\int_s^r f = -\int_r^s f$.

The operator $\int$ is

<u>Additive</u>: $\int_r^s (f + g) = \int_r^s f + \int_r^s g,$

<u>Homogeneous</u>: $\int_r^s (k \cdot f) = k \cdot \int_r^s f,$ k any constant.

The following result provides a way of expressing $\int_r^s f$ as the limit of a sequence of sums.

> If f is continuous on [r,s], then $\int_r^s f$ equals the limit of the convergent sequence whose nth term is the sum
>
> $$j_n \cdot f(a_1) + \cdots + j_n \cdot f(a_{n-1}),$$
>
> where $a_1,\ldots,a_n$ is the n-fold arithmetic partition of [r,s], and j_n is the common value of
>
> $$a_2 - a_1, \ldots, a_n - a_{n-1}.$$

The operator $\int$ is characterized by the following three properties:

1. For any interval [r,s] and any constant function k(x) = c on [r,s],

$$\int_r^s k = c \cdot (s - r).$$

2. For any interval [r,s] and any continuous functions f and g on [r,s], if $f(x) \le g(x)$ on [r,s], then

$$\int_r^s f \le \int_r^s g.$$

3. For any numbers r, s, t such that $r < s < t$ and any continuous function f on [r,t],

$$\int_r^s f + \int_s^t f = \int_r^t f.$$

1.10 THE FUNDAMENTAL THEOREMS OF CLASSICAL CALCULUS

The classical derivative and classical integral are related in the manner indicated by the following two theorems, the second of which is a simple consequence of the Basic Theorem of Classical Calculus.

<u>First Fundamental Theorem of Classical Calculus</u>

If f is continuous on [r,s], and

$$g(x) = \int_r^x f, \quad \text{for every number } x \text{ in } [r,s],$$

then

$$Dg = f, \quad \text{on the interior of } [r,s].$$

<u>Second Fundamental Theorem of Classical Calculus</u>

If Dh is continuous on [r,s], then

$$\int_r^s (Dh) = h(s) - h(r).$$

CHAPTER TWO

Meta – Calculus

2.1 INTRODUCTION

The classical gradient of a function f on an interval [r,
depends only on TWO points of f, namely (r, f(r)) and (s, f(s)
If f is suitably restricted, we shall show how to define new gradients that depend on ALL the points (x, f(x)) for which $r \le x \le s$. We call each such gradient a meta-gradient and eac
resulting calculus a meta-calculus.

We are using the prefix "meta" because each meta-calculus transcends the classical calculus in the manner explained abov
and in another way that will appear in Section 2.13.

2.2 WEIGHT FUNCTIONS

A <u>weight function</u> is any function that is continuous and positive on R. Of course there are infinitely-many weight functions.

Each pair of weight functions determines a meta-calculus, of which there are infinitely-many.

Henceforth u and v are arbitrarily selected weight functions.

2.3 META-MEASURE OF AN INTERVAL

To facilitate comparisons between classical calculus and meta-calculus we shall use the symbol m[r,s] for s - r, the classical measure of [r,s].

First observe that m[r,s] equals the positive limit of the constant (and hence convergent) sequence whose nth term is the sum

$$1\cdot(a_2 - a_1) + \cdots + 1\cdot(a_n - a_{n-1}),$$

where $a_1,\ldots,a_n$ is the n-fold arithmetic partition of [r,s]. This suggests the following definition.

The <u>meta-measure of an interval [r,s]</u> is the positive limit of the convergent sequence whose nth term is the sum

$$u(a_1)\cdot(a_2 - a_1) + \cdots + u(a_{n-1})\cdot(a_n - a_{n-1}),$$

where $a_1,\ldots,a_n$ is the n-fold arithmetic partition of [r,s].

We shall use the symbol $\ddot{m}$[r,s] for the meta-measure of [r,s]. Clearly

$$\ddot{m}[r,s] = \int_r^s u,$$

which should be contrasted with

$$m[r,s] = \int_r^s 1.$$

Of course, if $u(x) = 1$ on [r,s], then

$$\ddot{m}[r,s] = m[r,s].$$

Finally, if $r < t < s$, then

$$\ddot{m}[r,t] + \ddot{m}[t,s] = \ddot{m}[r,s].$$

2.4 META-CHANGE OF A FUNCTION

To facilitate comparisons we shall use the symbol $C_r^s f$ for $f(s) - f(r)$, the classical change of f on [r,s].

Now observe that $C_r^s f$ is equal to the limit of the constant (and hence convergent) sequence whose nth term is the sum

$$1\cdot[f(a_2) - f(a_1)] + \cdots + 1\cdot[f(a_n) - f(a_{n-1})],$$

where $a_1,\ldots,a_n$ is the n-fold arithmetic partition of [r,s]. This suggests the following discussion.

If f is a smooth function on [r,s], then one can establish the convergence of the sequence whose nth term is the sum

$$\upsilon(a_1)\cdot[f(a_2) - f(a_1)] + \cdots + \upsilon(a_{n-1})\cdot[f(a_n) - f(a_{n-1})],$$

where $a_1,\ldots,a_n$ is the n-fold arithmetic partition of [r,s]. Accordingly, we define the <u>meta-change of a smooth function f on [r,s]</u> to be the limit of the foregoing sequence and we denote that limit by $\ddot{C}_r^s f$.

Notice that $\ddot{C}_r^s f$ has been defined only if f is a smooth function on [r,s].

It turns out that

$$\ddot{C}_r^s f = \int_r^s (\upsilon \cdot Df),$$

which should be contrasted with

$$C_r^s f = \int_r^s (1 \cdot Df).$$

Also, $\ddot{C}_r^s f$ equals the Stieltjes integral of υ relative to f on [r,s].

The operator $\overset{\prime\prime}{C}$ is

<u>Additive</u>: $\overset{\prime\prime}{C}{}_r^s(f + g) = \overset{\prime\prime}{C}{}_r^s f + \overset{\prime\prime}{C}{}_r^s g$,

<u>Homogeneous</u>: $\overset{\prime\prime}{C}{}_r^s(k \cdot f) = k \cdot \overset{\prime\prime}{C}{}_r^s f$, k any constant.

As expected, if f is constant on [r,s], then $\overset{\prime\prime}{C}{}_r^s f = 0$. And, of course, if $v(x) = 1$ on [r,s], then

$$\overset{\prime\prime}{C}{}_r^s f = C_r^s f.$$

Finally, if $r < t < s$, then

$$\overset{\prime\prime}{C}{}_r^t f + \overset{\prime\prime}{C}{}_t^s f = \overset{\prime\prime}{C}{}_r^s f.$$

2.5 META-UNIFORM FUNCTIONS

We know that the classical change of each given linear function is the same on any two intervals with the same classical measure. This suggests that we seek a class of functions having a corresponding property with respect to meta-change and meta-measure.

First we define a <u>meta-unit interval</u> to be any interval whose meta-measure is 1. There may or may not exist meta-unit intervals, depending on the weight function u.

Henceforth we let

$$W(x) = \int_0^x (u/v), \quad \text{for each number } x.$$

Now consider any function j such that

$$j(x) = b \cdot W(x) + c, \quad \text{on } R,$$

where b and c are constants. Clearly j is a smooth function on every interval. It turns out that the meta-change of j is the same on any two intervals with the same meta-measure. Therefore each function on R that is expressible in the form $b \cdot W(x) + c$, and only such a function, will be called a meta-uniform function

In particular, each meta-uniform function has the same meta change on every meta-unit interval, if any.

Clearly every constant function on R is meta-uniform (take b = 0).

Moreover, if u/v is constant, then the meta-uniform functions are identical with the linear functions.

2.6 META-SLOPE OF A META-UNIFORM FUNCTION

We know that if ℓ is a linear function, then on EVERY interval [r,s] the ratio

$$\frac{C_r^s \ell}{m[r,s]}$$

has the same value, which we call the classical slope of ℓ.

Similarly, if j is a meta-uniform function, then one can easily prove that on EVERY interval [r,s] the ratio

$$\frac{\overset{\shortparallel}{C}{}_r^s j}{\overset{\shortparallel}{m}[r,s]}$$

has the same value, which we call the meta-slope of j.

Of course the meta-slope of a meta-uniform function is equal to its meta-change on every meta-unit interval, if any.

The meta-slope of the meta-uniform function $b \cdot W(x) + c$ turns out to be b.

2.7 META-GRADIENT

The classical gradient $G_r^s f$ of a function f on [r,s] was defined to be the ratio of the classical change of f on [r,s] to the classical measure of [r,s]; that is,

$$G_r^s f = \frac{C_r^s f}{m[r,s]} .$$

This suggests the following definition.

The <u>meta-gradient of a smooth function f on [r,s]</u> is denoted by $\ddot{G}_r^s f$ and is defined to be the ratio of the meta-change of f on [r,s] to the meta-measure of [r,s]; that is,

$$\ddot{G}_r^s f = \frac{\ddot{C}_r^s f}{\ddot{m}[r,s]} .$$

Since $\ddot{C}_r^s f$ depends on all the points (x, f(x)) for which $r \leq x \leq s$, it follows that $\ddot{G}_r^s f$ also depends on all those points. This is in sharp contrast to the classical gradient $G_r^s f$, which depends only on the two points (r, f(r)) and (s, f(s)).

It is worth noting that

$$\ddot{G}_r^s f = \frac{\int_r^s (v \cdot Df)}{\int_r^s u},$$

which should be contrasted with

$$G_r^s f = \frac{\int_r^s (1 \cdot Df)}{\int_r^s 1}$$

The operator $\ddot{G}$ is

<u>Additive</u>: $\ddot{G}_r^s(f + g) = \ddot{G}_r^s f + \ddot{G}_r^s g,$

<u>Homogeneous</u>: $\ddot{G}_r^s(k \cdot f) = k \cdot \ddot{G}_r^s f,$ k any constant.

If j is a meta-uniform function then $\ddot{G}_r^s j$ equals the meta-slope of j.

If u and v are constant and equal on [r,s], then

$$\ddot{G}_r^s f = G_r^s f.$$

Finally, when r = s, the expression for the meta-gradient yields the indeterminate form 0/0, which brings us to the next topic.

2.8 META-DERIVATIVE

Let f be a smooth function on an interval containing the number a in its interior. Notice that for each number x in that interval,

$$\frac{\int_a^x (v \cdot Df)}{\int_a^x u} = \begin{cases} \overset{''}{G}{}_a^x f & \text{if } a < x, \\ \overset{''}{G}{}_x^a f & \text{if } x < a. \end{cases}$$

We now define the <u>meta-derivative of f at a</u> to be the following limit, which necessarily exists,

$$\lim_{x \to a} \frac{\int_a^x (v \cdot Df)}{\int_a^x u},$$

and we denote that limit by $[\overset{''}{D}f](a)$.

Keep in mind that $[\overset{''}{D}f](a)$ has been defined only if f is a smooth function on an interval containing a in its interior.

It turns out that

$$[\overset{''}{D}f](a) = \frac{v(a)}{u(a)} \cdot [Df](a).$$

The <u>meta-derivative of f</u>, denoted by $\overset{''}{D}f$, is the function that assigns to each number t the number $[\overset{''}{D}f](t)$, if it is defined.

It is worth noting that if f is a smooth function on an interval [r,s], then $\overset{''}{D}f$ is continuous on the interior of [r,s].

The operator $\ddot{D}$ is

<u>Additive</u>: $[\ddot{D}(f + g)](a) = [\ddot{D}f](a) + [\ddot{D}g](a)$,

<u>Homogeneous</u>: $[\ddot{D}(k \cdot f)](a) = k \cdot [\ddot{D}f](a)$, k any constant.

As expected, if f is a constant function on R, then $\ddot{D}f$ is everywhere equal to 0.

Furthermore the meta-derivative of each meta-uniform function has a constant value equal to its meta-slope. Indeed, only meta-uniform functions have meta-derivatives that are constant on R.

In classical calculus, Dg = g if g(x) = exp x on R. In meta-calculus, $\ddot{D}g = g$ if $g(x) = \exp[W(x)]$ on R. (Recall that $W(x) = \int_0^x (u/v)$ for each number x.)

In the special case where $v(x) = 1$ on R, $[\ddot{D}f](a)$ turns out to be the classical derivative of f relative to U at a, where $U(x) = \int_0^x u$ for each number x:

$$[\ddot{D}f](a) = \frac{[Df](a)}{[DU](a)} = \lim_{x \to a} \frac{f(x) - f(a)}{U(x) - U(a)} .$$

Of course, if $u(a) = v(a)$, then

$$[\ddot{D}f](a) = [Df](a).$$

We conclude this section with a mean value theorem of meta-calculus:

If f is a smooth function on [r,s], then strictly between r and s there is a number c such that

$$[\ddot{D}f](c) = \ddot{G}_r^s f.$$

2.9 META-AVERAGE

Let f be any continuous function on [r,s]. Then the arithmetic average of f on [r,s] was denoted by $A_r^s f$ and was defined to be the limit of the convergent sequence whose nth term is the arithmetic average

$$\frac{f(a_1) + \cdots + f(a_n)}{n},$$

where $a_1, \ldots, a_n$ is the n-fold arithmetic partition of [r,s].

Our definition of the meta-average of a continuous function is based, not on arithmetic partitions, but on meta-partitions, which we define next.

A _meta-partition of an interval [r,s]_ is any finite sequence of numbers $b_1, \ldots, b_n$ such that

$$r = b_1 < b_2 < \cdots < b_n = s$$

and

$$\ddot{m}[b_1, b_2] = \cdots = \ddot{m}[b_{n-1}, b_n].$$

A meta-partition with exactly n terms is said to be _n-fold_.

Let f be any continuous function on [r,s]. Then the _meta-average of f on [r,s]_ is denoted by $\ddot{A}_r^s f$ and is defined to be the limit of the convergent sequence whose nth term is the arithmetic average

$$\frac{f(b_1) + \cdots + f(b_n)}{n},$$

where $b_1, \ldots, b_n$ is the n-fold meta-partition of [r,s].

Surprisingly it turns out that $\ddot{A}_r^s f$ equals the following weighted arithmetic average of f on [r,s]:

$$\frac{\int_r^s (u \cdot f)}{\int_r^s u}.$$

One can also obtain $\ddot{A}_r^s f$ as the limit of the convergent sequence whose nth term is the weighted arithmetic average

$$\frac{u(a_1) \cdot f(a_1) + \cdots + u(a_n) \cdot f(a_n)}{u(a_1) + \cdots + u(a_n)},$$

where $a_1,\ldots,a_n$ is the n-fold arithmetic partition of [r,s].

The operator $\ddot{A}$ is

<u>Additive</u>: $\ddot{A}_r^s(f + g) = \ddot{A}_r^s f + \ddot{A}_r^s g$,

<u>Homogeneous</u>: $\ddot{A}_r^s(k \cdot f) = k \cdot \ddot{A}_r^s f$, k any constant.

The operator $\ddot{A}$ is characterized by the following three properties:

1. For any interval [r,s] and any constant function k(x) = c on [r,s],

$$\ddot{A}_r^s k = c.$$

2. For any interval [r,s] and any continuous functions f and g on [r,s], if $f(x) \le g(x)$ on [r,s], then

$$\ddot{A}_r^s f \le \ddot{A}_r^s g.$$

3. For any numbers r, s, t such that $r < s < t$ and any continuous function f on [r,t],

$$\overset{''}{m}[r,s] \cdot \overset{''}{A}{}_r^s f + \overset{''}{m}[s,t] \cdot \overset{''}{A}{}_s^t f = \overset{''}{m}[r,t] \cdot \overset{''}{A}{}_r^t f.$$

If the weight function u is constant on [r,s], then

$$\overset{''}{A}{}_r^s f = A_r^s f.$$

We now state another mean value theorem of meta-calculus:

If f is continuous on [r,s], then strictly between r and s there is a number c such that

$$\overset{''}{A}{}_r^s f = f(c);$$

that is, the meta-average of a continuous function on [r,s] is assumed at some argument strictly between r and s.

The following restatement of the preceding theorem is an immediate consequence of the well-known Second Mean Value Theorem for Classical Integrals:

If f is continuous on [r,s], then strictly between r and s there is a number c such that

$$\int_r^s (f \cdot u) = f(c) \cdot \int_r^s u.$$

2.10 THE BASIC THEOREM OF META-CALCULUS

First we state a simple result about any smooth function h on [r,s].

Let $a_1,\ldots,a_n$ be any arithmetic partition of [r,s]. For each integer i from 1 to n-1, let

$$w_i = A_{a_i}^{a_{i+1}} u.$$

Then

$$\frac{w_1 \cdot \ddot{G}_{a_1}^{a_2} h + \cdots + w_{n-1} \cdot \ddot{G}_{a_{n-1}}^{a_n} h}{w_1 + \cdots + w_{n-1}} = \ddot{G}_r^s h;$$

that is, a certain weighted arithmetic average of the n-1 successive meta-gradients of h determined by a_1, ..., a_n is equal to the meta-gradient of h on [r,s].

The preceding result suggests the following important theorem.

<u>Basic Theorem of Meta-Calculus</u>

If $\ddot{D}h$ is defined on [r,s] (and hence is necessarily continuous on [r,s]), then its meta-average on [r,s] is equal to the meta-gradient of h on [r,s]; that is,

$$\ddot{A}_r^s(\ddot{D}h) = \ddot{G}_r^s h.$$

In view of the preceding theorem we say that the meta-average fits naturally into the scheme of meta-calculus.

<u>N O T E</u>

For each smooth function h on [r,s], we have another simple result.

Let $b_1,\ldots,b_n$ be any meta-partition of [r,s]. Then

$$\frac{\ddot{G}_{b_1}^{b_2} h + \cdots + \ddot{G}_{b_{n-1}}^{b_n} h}{n - 1} = \ddot{G}_r^s h;$$

that is, the unweighted arithmetic average of the n-1 successive meta-gradients of h determined by $b_1,\ldots,b_n$ is equal to the meta-gradient of h on [r,s].

2.11 THE BASIC PROBLEM OF META-CALCULUS

Suppose that f, the meta-derivative of h, is known at each number in [r,s]. Find $\ddot{C}_r^s h$.

<u>Solution</u>

By the Basic Theorem of Meta-Calculus,

$$\ddot{A}_r^s f = \ddot{G}_r^s h = \frac{\ddot{C}_r^s h}{\ddot{m}[r,s]}.$$

Hence,

$$\ddot{C}_r^s h = \ddot{m}[r,s] \cdot \ddot{A}_r^s f.$$

The number $\overset{''}{m}[r,s] \cdot \overset{''}{A}{}_r^s f$ arises with sufficient frequency to warrant a special name, "the meta-integral of f on [r,s]," which is discussed in the next section.

Thus, the Basic Theorem of Meta-Calculus, which involves the meta-average, meta-derivative, and meta-gradient, provides for the Basic Problem of Meta-Calculus an immediate solution, which, in turn, motivates our definition of the meta-integral in terms of the meta-average.

2.12 META-INTEGRAL

Let f be any continuous function on [r,s]. Then the <u>meta-integral of f on [r,s]</u> is denoted by $\overset{''}{\int}_r^s f$ and is defined to be the number

$$\overset{''}{m}[r,s] \cdot \overset{''}{A}{}_r^s f.$$

We set $\overset{''}{\int}_r^r f = 0$.

It turns out that

$$\overset{''}{\int}_r^s f = \int_r^s (u \cdot f),$$

which is called an inner product in certain contexts.

The operator $\overset{''}{\int}$ is

<u>Additive</u>: $\overset{''}{\int}_r^s (f + g) = \overset{''}{\int}_r^s f + \overset{''}{\int}_r^s g$,

<u>Homogeneous</u>: $\overset{''}{\int}_r^s (k \cdot f) = k \cdot \overset{''}{\int}_r^s f$, k any constant.

The following results provide two ways of expressing $\overset{''}{\int}_r^s f$ as the limit of a sequence of sums.

1. If f is continuous on [r,s], then $\overset{''}{\int}_r^s f$ equals the limit of the convergent sequence whose nth term is the sum

$$j_n \cdot u(a_1) \cdot f(a_1) + \cdots + j_n \cdot u(a_{n-1}) \cdot f(a_{n-1}),$$

where $a_1, \ldots, a_n$ is the n-fold arithmetic partition of [r,s], and j_n is the common value of

$$m[a_1, a_2], \ldots, m[a_{n-1}, a_n].$$

2. If f is continuous on [r,s], then $\overset{''}{\int}_r^s f$ equals the limit of the convergent sequence whose nth term is the sum

$$k_n \cdot f(b_1) + \cdots + k_n \cdot f(b_{n-1}),$$

where $b_1, \ldots, b_n$ is the n-fold meta-partition of [r,s], and k_n is the common value of

$$\overset{''}{m}[b_1, b_2], \ldots, \overset{''}{m}[b_{n-1}, b_n].$$

Since $k_n = U(b_{i+1}) - U(b_i)$ for $i = 1, \ldots, n-1$, where $U(x) = \int_0^x u$ for each number x, it is not surprising that $\overset{''}{\int}_r^s f$ is the Stieltjes integral of f relative to U on [r,s].

The operator $\overset{''}{\int}$ is characterized by the following three properties:

1. For any interval [r,s] and any constant function $k(x) = c$ on [r,s],

$$\overset{''}{\int}_r^s k = c \cdot \overset{''}{m}[r,s].$$

2. For any interval [r,s] and any continuous functions f and g on [r,s], if f(x) ≤ g(x) on [r,s], then

$$\int_r^{''s} f \le \int_r^{''s} g.$$

3. For any numbers r, s, t such that r < s < t and any continuous function f on [r,t],

$$\int_r^{''s} f + \int_s^{''t} f = \int_r^{''t} f.$$

We conclude this section by noting that if $u(x) = 1$ on [r,s], then

$$\int_r^{''s} f = \int_r^{s} f.$$

2.13 THE FUNDAMENTAL THEOREMS OF META-CALCULUS

The meta-derivative and meta-integral are related in the manner indicated by the following two theorems, the second of which is a simple consequence of the Basic Theorem of Meta-Calculus.

<u>First Fundamental Theorem of Meta-Calculus</u>

If f is continuous on [r,s], and

$$g(x) = \int_r^{''x} f, \quad \text{for every number x in [r,s],}$$

then

$$\overset{''}{D}g = \upsilon \cdot f, \quad \text{on the interior of [r,s].}$$

Notice that the right-hand side of the preceding equation is $\upsilon \cdot f$, in contrast to the First Fundamental Theorem of Classical Calculus, in which the right-hand side of the corresponding equation is f. This distinction provides another reason for using the prefix "meta."

<u>Second Fundamental Theorem of Meta-Calculus</u>

If $\ddot{D}h$ is defined on [r,s] (and hence is necessarily continuous on [r,s]), then

$$\ddot{\int}_r^s (\ddot{D}h) = \ddot{C}_r^s h.$$

Notice that the preceding equation has exactly the same form as the equation in the Second Fundamental Theorem of Classical Calculus.

O C A C M N N C

LIST OF SYMBOLS

INDEX

BIBLIOGRAPHY

Grossman, M. and Katz, R. *Non-Newtonian Calculus*. Rockport, MA: Mathco, 1972.
Included in this book, which was the first publication on non-Newtonian calculus, are discussions of nine specific non-Newtonian calculi, the general theory of non-Newtonian calculus, and heuristic guides for the application thereof.

Grossman, M. *The First Nonlinear System of Differential and Integral Calculus*. Rockport, MA: Mathco, 1979.
This book contains a detailed account of the geometric calculus, which was the first of the non-Newtonian calculi. Also included are discussions of the analogy that led to the discovery of that calculus, and some heuristic guides for its application.

Meginniss, J. R. "Non-Newtonian Calculus Applied to Probability, Utility, and Bayesian Analysis." *Proceedings of the American Statistical Association*: Business and Economics Statistics Section (1980), pp. 405-410.
This paper presents a new theory of probability suitable for the analysis of human behavior and decision making. The theory is based on the idea that subjective probability is governed by the laws of a non-Newtonian calculus and one of its corresponding arithmetics.

Grossman, J., Grossman, M., and Katz, R. *The First Systems of Weighted Differential and Integral Calculus*. Rockport, MA: Archimedes Foundation, 1980.
This monograph reveals how weighted averages, Stieltjes integrals, and derivatives of one function with respect to another can be linked to form systems of calculus, which are called weighted calculi because in each such system a weight function plays a central role.

Grossman, M. *Bigeometric Calculus: A System with a Scale-Free Derivative*. Rockport, MA: Archimedes Foundation, 1983.
This book contains a detailed treatment of the bigeometric calculus, which has a derivative that is scale-free, i.e., invariant under all changes of scales (or units) in function arguments and values. Also included are heuristic guides for the application of that calculus, and various related matters such as the bigeometric method of least squares.

Grossman, J., Grossman, M., and Katz, R. *Averages: A New Approach*. Rockport, MA: Archimedes Foundation, 1983.
This monograph is primarily concerned with a comprehensive family of averages (unweighted and weighted) of functions that arose naturally in the development of non-Newtonian calculus and weighted non-Newtonian calculus. The monograph also contains discussions of some heuristic guides for the appropriate use of averages and an interesting family of means of two positive numbers.

ISBN 0977117022

www.ingramcontent.com/pod-product-compliance
Lightning Source LLC
LaVergne TN
LVHW050948080826
845145LV00004B/1454

* 9 7 8 0 9 7 7 1 1 7 0 2 4 *